By Bonnie Beers

Table of Contents

Consulting Editor: Gail Saunders-Smith, Ph.D.
Consultants: Claudine Jellison and
Patricia Williams, Reading Recovery Teachers
Content Consultant: Barbara Benson,
Ph.D. of History, Executive Director
of the Historical Society of Delaware

The Earth is round like a ball.
Most of the Earth is covered
by water.
The rest is land.

If the Earth's surface
were a pizza with four slices,
about three slices would be water
and one slice would be land.

Most of the water on Earth
is salt water.
Oceans and seas have salt water.
Oceans are bigger than seas.

Land near an ocean or sea
is called the coast.
Some coasts are rocky.
Some coasts are sandy.

Rivers are long bodies
of fresh water.
Sometimes a river moves fast.
Sometimes a river moves slow.

Lakes are bodies of fresh water that have land all around them.

Ponds are like lakes, but smaller.
Ponds have land
all around them too.

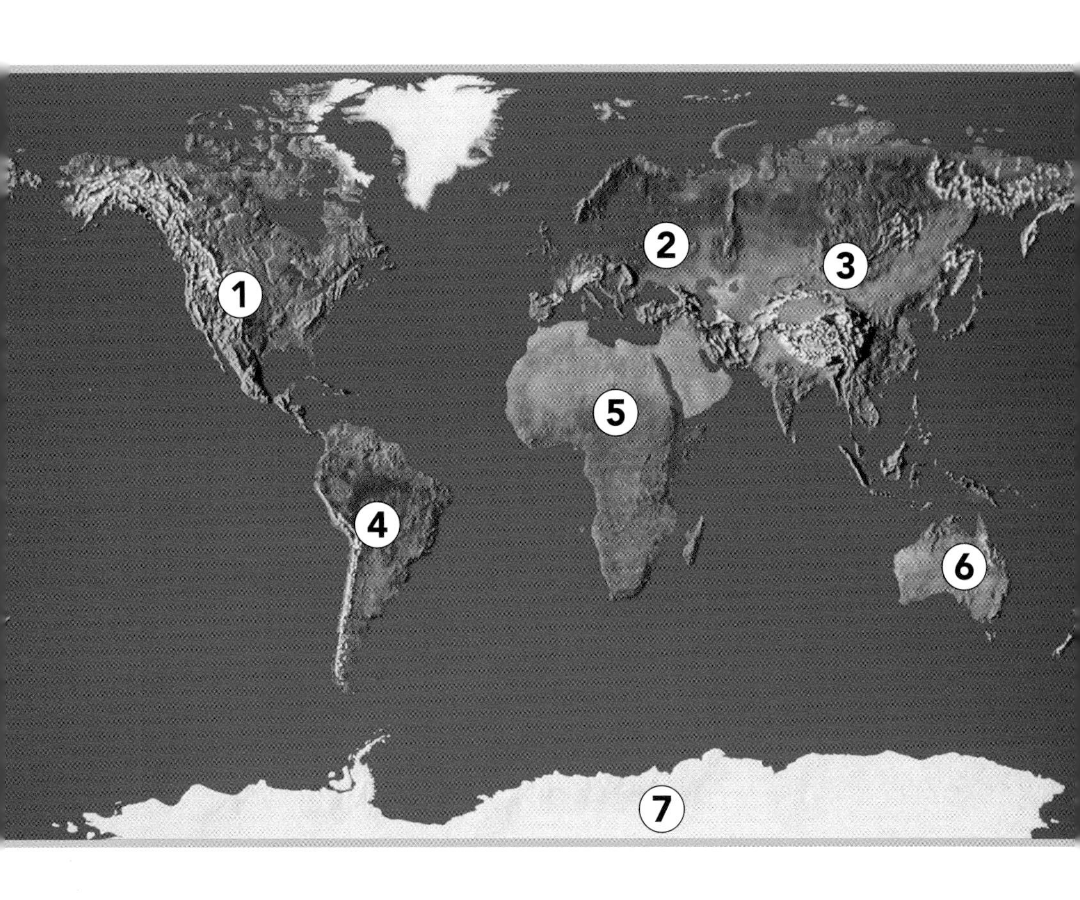

Land has many shapes and sizes. The seven biggest bodies of land are called continents.

An island is land
that has water all around it.

Mountains are very high land.
Hills are land that is not as high.

A valley is low land between some mountains and hills. Streams flow through some valleys.

Plains are wide flat lands.
Farmers often plant crops on plains.

Some large areas
of land are forests.
They are covered
with trees and plants.
There are different kinds of forests.

Land that is very wet
is called a wetland.
Land that is very dry
is called a desert.

Earth is made up of many types of land and water.